YOUR KNOWLEDGE HAS VALUE

- We will publish your bachelor's and
  master's thesis, essays and papers

- Your own eBook and book -
  sold worldwide in all relevant shops

- Earn money with each sale

Upload your text at www.GRIN.com
and publish for free

Lithium Extraction between Neo-Extractivism and Sustainability. The Case of Bolivia

Joost Zickler

Bibliographic information published by the German National Library:

The German National Library lists this publication in the National Bibliography; detailed bibliographic data are available on the Internet at http://dnb.dnb.de.

ISBN: 9783389011133
This book is also available as an ebook.

© GRIN Publishing GmbH
Trappentreustraße 1
80339 München

Print and binding: Books on Demand GmbH, Norderstedt, Germany
Printed on acid-free paper from responsible sources.

The present work has been carefully prepared. Nevertheless, authors and publishers do not incur liability for the correctness of information, notes, links and advice as well as any printing errors.

GRIN web shop: https://www.grin.com/document/1461288

Geographical Economics: Regional Strategies and Sustainability

Seminar Paper

Lithium extraction between neo-extractivism and sustainability:

The case of Bolivia

Presented by Joost Zickler

University of Cologne

VWL sozialwiss. Richtung

Faculty of Management, Economics and Social Sciences

Department of Economic and Social Geography

29/01/2024

Table of Contents

List of abbreviations

DLE – Direct Lithium Extraction

GPN – Global Production Network

MAS – Movimiento Al Socialismo

MHE – Ministerio de Hidrocarburos y Energías *(Ministry of Hydrocarbons and Energies)*

NE – Neo-Extractivism

YLB – Yacimientos de Litio Boliviano

List of Figures

All translations from Spanish to English have been conducted with the help of the software DeepL (https://www.deepl.com/translator) and revised by the author.

1. Introduction

"The industrialization of the Salar de Uyuni is declared a national priority for the productive, economic and social development of the Department of Potosi." (Decreto Supremo de Bolivia No. 29496, 2008).

With these words, the Bolivian government launched the industrialization of Bolivian lithium in 2008.

Bolivia possesses the world's largest lithium reserves (Sanchez-Lopez, 2019: 1). Lithium is considered one of the most important minerals for the global energy transition, given its key role in battery production, especially for electric vehicles (Heredia et al, 2020: 213ff). The deep wounds of Bolivia's colonial past cause a strong alertness towards resource extraction in the country. This makes the Bolivian lithium decision such an interesting case: It was Evo Morales, Bolivia's first indigenous president, who launched the lithium extraction.

Resource extraction policies of progressive Latin American governments are at the core of a literature centered around the concept of "neo-extractivism" (NE) introduced by Gudynas (2009). NE scholars seek to unearth the reasons that make progressive countries exploit their natural resources in a way that damages them environmentally and socially.

In this paper, I want to link the Bolivian lithium extraction policies to NE. Following a qualitative process-tracing approach, I wish to answer the question: Are the Bolivian lithium policies since 2008 on the verge of continuing neo-extractivism?

I start by reviewing the relevant literature around extractivism in general and NE in particular, elaborating five defining hallmarks of NE. I also investigate the relation of NE and sustainability. In section three, I deep-dive into Bolivian lithium extraction policies, based on government communication data, presidential decrees and the Bolivian constitution, complemented by academic literature. In section 4, based on these insights, I answer the research question, checking for the criteria from section 3. I also address conceptual shortcomings of the NE concept with insights from the global lithium production network literature. In section 5, I will conclude the paper, assessing the results from a sustainability perspective.

2. Literature review

2.1. Extractivism and sustainability

Generations of scholars referred to the concept of "extractivism" to describe aa economic development model that is based on the extraction of natural resources. According to Gudynas (2013: 5), "extractivism is a particular case of intense or high-volume extraction of natural resources for export, with no or limited processing." A very large body of literature that has emerged around extractivism proves the concept's empirical relevance.

Scholars soon recognized that there are "variations of extractivism" (Bruna, 2022: 842) which follow the same basic pattern but which differ on some dimensions. Some focus on "extractive imperialism" during colonialism and root extractivism in the 15th century (Girvan, 2014: 49f). Others focus on modern forms of extractivism: Bruna (2022: 842) differentiates between "mining" and "agrarian" extractivism. Dunlap and Jakobsen (2019: 94ff) introduce the concept of "green extractivism", pointing out that increasingly, institutions as governments or the World Bank seek to legitimize extractivist activities presenting them as necessary to fight climate change, referring to the extraction of minerals needed for new energy technologies (Andreucci et al, 2023: 3). Gudynas (2012: 131) himself introduces the notion of the neoliberal "classical extractivism" of the 1980s/90s that was prevalent in Latin America and dominated by powerful transnational enterprises.

A common feature of all variations of extractivism is the lack of sustainability. Sustainable development, as defined in the UN Brundtland Report from 1987, is "development that meets the needs of the present without compromising the ability of future generations to meet their own needs." (World Commission on Environment and Development, 1987). Hence, it has not merely an environmental, but also an economic and social dimension.

Only few papers explicitly state that extractivism is not sustainable, but it is apparent that the key features of extractivism are inconsistent with sustainability. On an environmental dimension, extractivism undermines its own foundations by exploiting finite natural resources. Also, extractivism is linked to environmental damage in various forms, which in addition cause social conflicts or even human rights violations and destroy the fundament of a stable society (Gudynas, 2009: 204).

Also economically, extractivism is not sustainable as it does not build foundations for a prospering economy for future generations. According to the infant industry argument forwarded by Hamilton (2017 [1790]) and List (2007 [1841]), the exposition to a very competitive import sector avoids that domestic young industries can experience learning effects that lead to efficiency gains and competitiveness. And a competitive import sector is a consequence of extractivism, because extractivist exportations lead to foreign currency inflows that facilitate low-cost imports. Hence, consumption needs are being met by importations and the demand for domestic products is low, so that the domestic industry cannot develop. Future generations therefore stay dependant on imports and currency inflows (see also, Gudynas, 2012: 139).

2.2. Neo-extractivism and sustainability

Another important variety of extractivism is "new" or "neo-extractivism" (NE) as introduced by Gudynas (2009, also 2012, 2013, 2015). Since then, his following conceptualization can be considered state-of-the-art. Hardly any NE paper does not refer to Gudynas.

Gudynas (2009: 189) observes a comparable development in various Latin American countries including Bolivia, Chile, Argentina, Ecuador, Brazil, Uruguay, and Venezuela in the early 2000s: The democratic emergence of progressive and/or left-wing governments after the period of classical extractivism.

Those governments promised to set an end to the neoliberal economic exploitation by strengthening the role of the state as a central as a sovereign planner that is in control of the countries' natural resources (ibid: 188ff). Strikingly, those governments stuck to the extraction of natural resources and even intensified it (ibid: 190). But the logic and framing changed. With a new national self-consciousness, they started to declare natural resources as national gifts that should benefit the entire population (as will be discussed below). They restricted the repatriation of profits by transnational firms and started to use extraction export profits for social spending (Obaya, 2021: 208f).

However, NE is still not sustainable. On an environmental level, the finiteness of the extracted resources persists. Also, in case studies, scholars point out that environmental contamination, loss of biodiversity and other environmental issues remain problematic under NE (for Bolivia, see Rodriguez Fernandez, 2020: 32f or Fabricant/Postero, 2019). Those same environmental issues also reflect in the emergence of new social conflicts, in the form of, for example, suppression of Indigenous rights or even human rights violations (for Bolivia, see ibid), so that the foundation for a stable society is being damaged. Therefore, Gudynas (2009: 204f) includes the continuity of environmental degradation and emergence of social conflicts into the definition of NE.

Economically, the extractivist states now invest the yields into social spending and education, thereby potentially building the base for a prospering economy. But still, a productive domestic economy cannot develop as long as the extracted resources keep being exported with no or minimal processing and a major share of gross domestic product continues to be generated selling raw materials. Like under every extractivism, this stays a key feature of NE (Gudynas, 2009: 194, 196). Therefore, NE can on neither dimension be considered a sustainable development model.

The hallmarks of NE can be summarized as follows:

i. Raw materials are being extracted in the global south.

ii. This has negative environmental and social impacts for local communities.

iii. Those raw materials are being exported to the global north with no or minimal processing.

iv. The state plays an active role, capturing a major share of export yields.

v. The state uses those yields for social spending, to benefit the population.

3. Lithium extraction in Bolivia: YLB-ACISA

Editor's note: this figure was removed due to copyright issues.

Fig. 1: Map of the Uyuni salt flat. As can be seen, the flat is located in the Bolivian Andean highlands. (Source: Enviro Map. Downloaded from: https://enviro-map.com/wp-content/uploads/2009/02/Salar-de-Uyuni-map-700.jpg on January 29, 2024)

3.1. Context

With its reserves in the salt flat of Uyuni, Bolivia possesses the world's largest lithium reserves (Sanchez-Lopez, 2019: 1) (see Fig. 1). Enclosed under a solid salt surface, the lithium is in a liquid brine. Conventional extraction pumps this brine to the surface and lets the contained water evaporate in huge pools, so that the (still not pure) lithium can be further processed (see Fig. 2). This massively dehydrates the soils and needs lots of space (Ardila García, 2023: 189). Due to its importance, the lithium price has risen by 400% only from 2015 to 2018 (Heredia et al, 2020: 219). Therefore, the Bolivian lithium resources have a lot of potential economic weight. This fuels debates within Bolivian politics about the handling of this situation.

Editor's note: this figure was removed due to copyright issues.

Fig. 2: Lithium evaporation pools in the Bolivian highlands. As can be seen, the evaporation requires a lot of space and water. (Source: Website of the Bolivian Ministry of Hydrocarbons and Energies. Downloaded from: https://www.mhe.gob.bo/wp-content/uploads/2021/02/61848285_1266015520226668_902137913221840896_o-1536x807.jpg on January 29, 2024)

Since 2005, Bolivia has – with an interruption in 2020 – its first indigenous government, the socialist MAS party that is the political representation of Bolivia's indigenous population (Parsons, 2013: 12f). Under the first MAS president, Evo Morales Ayma, Bolivia reconstituted itself as the Pluri-National State of Bolivia (Estado Plurinacional de Bolivia). The new constitution from 2009 put an end to persistent legal discrimination of the indigenous majority and to a neo-liberal economy that served foreign economic interests in the first place.

The new constitution includes a large section about Bolivia's natural resources. In article 348 II, it states that "Natural resources are strategic and of public interest for the country's development." Moreover, "The State, through public, social or community entities, shall assume control and direction over the exploration, extraction, industrialization, transportation and commercialization of natural resources." (Art. 351 I) Also, "The Bolivian people shall have equitable access to the benefits derived from the extraction of all natural resources." (Art. 353) These passages prove the new extractivist logic.

3.2. Lithium Politics under the MAS administration

The Morales administration addressed the lithium question in 2008 with Supreme Decree No. 29496, 2018, deciding to make use of its resources in a sovereign way that would benefit the Bolivian population in the first place. As Morales testified before the Bolivian parliament in 2007: "All natural resources must pass into the hands of the Bolivian people, into the hands of the Bolivian State. [...] And it is not a question of nationalizing for the sake of nationalizing [...], we have the obligation to industrialize them." (Svampa/Stefanoni, 2007: 239f) Still, the current president Arce (also MAS) states that he wants to attract international companies to "benefit us with technology, investments and, above all, with a greater production of lithium, which is what we are betting on as the central axis for the economic development of our country." (Ministerio de Hidrocarburos y Energías – MHE, 2023a: np).

Bohoslavski (2020) examines in detail the achievements of social programs under Morales. Some examples illustrate how extraction profits are used in practice to benefit the population: between 2006 and 2019, public investments in "education, health, and infrastructure such as roads, electricity, water, and sanitation facilities" (Bohoslavski, 2020: 129) helped to half school dropout rate and to significantly improve health indicators such as infant and maternal mortality (ibid: 130). Social transfers such as the Renta Dignidad ("dignity pension") roughly halved the poverty and extreme poverty rate, thereby reducing the Gini coefficient for income inequality by 26% (ibid: 130). Also, in the specific context of lithium, the government installed a new scientific research centre to promote scientific advancement around lithium extraction and battery production (MHE, 2022: 17).

As Arce's latter statement shows, the problem for the lithium extraction is the lack of technical know-how within Bolivian engineering (see also, YLB, 2020: 1): Although YLB did make considerable progress in establishing extraction infrastructure and in developing extraction methods, it was unable to elevate the production to industrial scale (Obaya, 2021: 5).

So, in 2016, the government started to search for international partner firms that could help YLB advance its technological know-how and scale up its productivity (YLB, 2017: np). According to the constitution, the cooperation had to be under Bolivian control and YLB majority ownership (Obaya, 2021: 5). Also, the foreign partners were not allowed to pump brine from the salt flat (ibid: 5).

The German medium-size enterprise ACISA won the bidding (ibid: 5). ACISA was specialized on lithium extraction from residual brine, a waste product of YLB's lithium extraction via evaporation pools. Hence, the deal was framed with a supposed water neutrality (ibid: 5). This was of preliminary importance because the fear of freshwater deprivation was the biggest social and environmental risk related to the water-intensive lithium extraction in the Bolivian Andean highlands where water has always been scarce (Heredia et al, 2020: 224).

The terms and conditions of the cooperation between YLB and ACISA were manifested in the 2018 Supreme Decree No. 3738, 2018 by the Morales administration. As summarized also in Obaya (2021: 5f), the partnership was designed in the shape of two newly founded public-private enterprises. First, the YLB-ACISA stock corporation with the mandate to produce lithium compounds for the domestic and international market using only residual brine bought from YLB, and second, another firm with the mandate to produce batteries, which has never been founded. YLB-ACISA was owned 49% by ACISA as minority shareholder and 51% by the Bolivian state via YLB. Hence, 51% of the net profits were to go the Bolivian state, 49% to ACISA. ACISA was to provide its technological know-how and share it with YLB. Also, ACISA was to guarantee a stable access to the European lithium markets. This YLB-ACISA business model was to exist for 70 years (for all, Obaya, 2021: 5f or Supreme Decree No. 3738, 2018).

But the Bolivian government surprisingly cancelled the deal by Supreme Decree No. 4070 in 2019. The reason was resistance from the local indigenous committee COMCIPO from Potosí, where the Uyuni salt flat is located. Notwithstanding guarantees of minimal environmental burden and water neutrality, the community feared new environmental problems (as YLB would keep extracting lithium via water-intense evaporation pools) and did additionally not agree on several terms of the contract, most significantly that merely 3% of YLB-ACISA's yields were to be paid as royalties to the federal state of Potosí and that supposedly the terms would benefit

ACISA more than Bolivia (Obaya, 2021: 8). The argument between COMPICO and the federal Government can transparently be traced in a government response document (MHE, 2019: np).

In 2021, YLB resumed negotiations with ACISA, but so far, they have not agreed on new terms. Instead, YLB started another international call for applications in 2021, this time for companies to cooperate in the deployment of the new technique of "direct lithium extraction" (DLE), which allows for lithium extraction without water-intensive evaporation pools (YLB, 2024: 2). Eventually, Bolivia started 2-year pilot projects with a Russian firm and a Chinese consortium in 2024 (ibid). In 2023, YLB signed a deal about the construction of a pilot battery manufacturing plant with the Indian company ALTMIN (YLB, 2023a: np). A key feature of the deal is a technology transfer (YLB, 2023b: np). For these cooperations, the terms and conditions are less transparent, but all cooperations are still under the condition that Bolivia stays in control along all value creation steps including battery manufacturing (MHE, 2023b: np). In 2024, YLB started another international call for cooperations (MHE, 2024: np), which proofs Bolivia's aspirations to use competitional pressure to yield the most favouring conditions for itself.

4. Results: Lithium extraction between neo-extractivism and sustainability

We can now answer the question: Are the Bolivian lithium policies since 2008 on the verge of continuing neo-extractivism? I take up the hallmarks of NE from section 2.2, checking the criteria one by one.

i.   *Raw materials are being extracted in the global south.*

Yes, this is what is about to happen in the Bolivian case.

ii.   *This has negative environmental and social impacts for local communities.*

This is harder to assess, but likely, this is not the case. The cancellation of the YLB-ACISA contract due to indigenous protests proofs that the government takes environmental and social concerns of the affected local communities seriously. Moreover, the Bolivian constitution sets very narrow limitations for extractive activities. The Bolivian concern for environmental and social compatibility of lithium extraction reflects also in the insistence on the least invasive extraction technique. With DLE, the water issue might seriously be solved.

iii.   *Those raw materials are being exported to the global north with no or minimal processing.*

Here, the problem is the missing specification of "no or minimal processing" within NE literature. This vagueness leaves it to the authors' discretion to speak or not of NE. This makes the concept vulnerable to normative abuse, undermining its ability to draw normative conclusions from it in a transparent way. Luckily, we can get remedy from the literature on global production networks (GPNs) that traces concrete steps of value creation.

There is even a GPN literature dealing specifically with lithium: Bos and Forget (2021). They specify five lithium value creation steps: extraction, refining, processing/manufacturing, consumption and recycling (Bos and Forget, 2021: 17). This differentiation helps us to check our third criterion. I argue that at least steps 1-3 should take place in the extracting country to transcend exportation with "no or minimal processing", because for the development opportunities of domestic productivity, it is secondary what goods are being consumed. What counts, is that value creation takes place domestically. Therefore, if processing/manufacturing takes place domestically, the neo-extractivist logic is overcome.

As outlined, the deal with ACISA would have involved the Bolivian state in lithium extraction, refining and manufacturing by establishing a national battery industry. This is also the case for the latest pilot projects with Chinese, Russian and Indian enterprises. Hence, the third criterion is also not fulfilled.

iv.   The state plays an active role, capturing a major share of export yields.

Yes, this is the case, as discussed in detail in section 3.

v.   The state uses those yields for social spending, to benefit the population.

This is also the case, as elaborated on the basis of the statements and examples in section 3.

5. Conclusion: A way towards sustainability

Stating that the Bolivian lithium extraction was just another case of neo-extractivism would not do justice to the efforts of the Bolivian government to avoid some fundamental problems of NE. This is the result of my qualitative analysis in which I traced Bolivian lithium policies since 2008. Those policies do not meet the defining hallmarks of NE as elaborated in section 2.2. The analysis contributes to NE literature by refining the concept of NE, linking it to the GPN literature and showing that NE is no longer applicable for all of its original coverage.

Notwithstanding the continuity of resource extraction, the Bolivian government proofs capable to emancipate from extractivism's distortions by, firstly, enclosing transnationals' power and deliberately profiting of selected benefits that transnationals can bring (e.g., funds and know-how); by secondly, actively focussing on the development of a domestic productive industry, and by thirdly, taking seriously social and environmental concerns of local and indigenous communities.

Therefore, with its lithium policy, Bolivia is taking serious steps toward a sustainable development model. One might argue that resource extraction can inherently never be sustainable (as it relies always on finite resources) and that therefore, the Bolivian lithium policies are not sustainable, but as extraction continues to be an inevitable reality of our global capitalist economic system, this perspective would not provide any new insights.

Nevertheless, it remains to be seen if Bolivia sticks to these high aspirations. But as long as there is no change of government, Bolivia's ongoing search for the best partner gives no reason to question that. Only the lack of transparency of the details around the new cooperations with Russian, Chinese and Indian state companies might be a reason for concern. Moreover, scholars should assess in detail the environmental burden of DLE. For reasons of capacity, this here paper could not deep-dive into that.

(3025 words without references)

6. List of references

Andreucci, D. et al. (2023). The coloniality of green extractivism: Unearthing decarbonisation by dispossession through the case of nickel. In: *Political Geography*, 107, article 102997. https://doi.org/10.1016/j.polgeo.2023.102997

Ardila García, V. (2023). Retos y desafíos geoeconómicos de Bolivia para la explotación del litio. In: *Perspectivas en Inteligencia,* 15(24), 173-205. http://doi. org/10.47961/2145194X.656

Bohoslavsky, J. P. (2020). Development and Human Rights in Bolivia: Advances, Contradictions, and Challenges. In: *Latin American Policy*, 11, 126-147. https://doi.org/10.1111/lamp.12181

Bos, V., & Forget, M. (2021). Global Production Networks and the lithium industry: A Bolivian perspective. In: *Geoforum*, 125, 168–180. https://doi.org/10.1016/j.geoforum.2021.06.001

Bruna, N. (2022). A climate-smart world and the rise of Green Extractivism. In: *The Journal of Peasant Studies*, 49(4), 839-864, DOI: 10.1080/03066150.2022.2070482

Dunlap, A. & Jakobsen, J. (2020). The Violent Technologies of Extraction: Political ecology, critical agrarian studies and the capitalist worldeater. eBook. Cham: Springer International Publishing. https://doi.org/10.1007/978-3-030-26852-7

Fabricant, N. & Postero, N. (2019). Performing Indigeneity in Bolivia: The Struggle Over the TIPNIS. In: Vindal Ødegaard, C., Rivera Andía, J. (eds.) *Indigenous Life Projects and Extractivism. Approaches to Social Inequality and Difference.* 245-276. Cham: Palgrave Macmillan. https://doi.org/10.1007/978-3-319-93435-8_10

Girvan, N. (2014). Extractive Imperialism in Historical Perspective. In: Petras, J., & Veltmeyer, H. (24 Jul. 2014). *Extractive Imperialism in the Americas*, 49-61. Leiden: Brill. https://doi.org/10.1163/9789004268869.

Gudynas, E. (2009). Diez tesis urgentes sobre el nuevo extractivismo. In: Dávila, F. R. et al. (eds.) *Extracivismo, política y sociedad*, 187-225. Quito: CAAP/CLAES.

Gudynas, E. (2012). Estado compensador y nuevos extractivismos: Las ambivalencias del progresismo sudamericano. In: *Nueva Sociedad,* 237, 128-146. https://biblioteca.hegoa.ehu.eus/downloads/18895/%2Fsystem%2Fpdf%2F2845%2FEstado_compensador_y_nuevos_extractivismos.pdf [accessed Jan 23, 2024]

Gudynas, E. (2013). Extracciones, extractivismos y extrahecciones. Un marco conceptual sobre la apropiación de recursos naturales. In: Observatorio del Desarollo, 18. CLAES, Montevideo.

Gudynas, E. (2015). Extractivismos. Ecología, economía y política de un modo de entender el desarollo y la Naturaleza. First edition. Cochabamba: CEDIB. https://catedra-tse.foronacionalambiental.org.co/wp-content/uploads/2021/05/ExtractivismosEcologiaPolitica.pdf [accessed Jan 23, 2024]

Hamilton, A. (2017). *Report on Manufacturers*. Great Neck Publishing. Online: https://search.ebsco-host.com/login.aspx?direct=true&db=asn&AN=21212912&site=ehost-live [accessed Jan 28. 2024]

Heredia, F., Martinez, A. L., & Surraco Urtubey, V. (2020). The importance of lithium for achieving a low-carbon future: Overview of the lithium extraction in the 'Lithium Triangle'. *Journal of Energy & Natural Resources Law*, 38(3), 213–236. https://doi.org/10.1080/02646811.2020.1784565

List, F. (2007). The national system of political economy. In: Barma, N. H. and Vogel, S. K. (eds.) *The Political Economy Reader*. 63-86. London: Routledge.

Ministerio de Hidrocarburos y Energías (2019). Respuestas a la fundamentación de COMPICO para la abrogación del Decreto Supremo No. 3738. *Press release*. Online: https://www.ylb.gob.bo/archivos/notas_archivos/aclaracionrespuestasds3738.pdf [accessed Jan 25, 2024]

Ministerio de Hidrocarburos y Energías (2022). Mas Energías para Salir Adelante #7. *Press release.* Online: https://www.mhe.gob.bo/2022/11/28/separata-mas-energias-para-salir-adelante-7-2-anos-de-gestion/ [accessed Jan 25, 2024]

Ministerio de Hidrocarburos y Energías (2023a). Gobierno anuncia nueva convocatoria con el objetivo de atraer mayores inversiones para el litio. *Press release.* Online: https://www.mhe.gob.bo/2023/12/20/gobierno-anuncia-nueva-convocatoria-con-el-objetivo-de-atraer-mayores-inversiones-para-el-litio/ [accessed Jan 23, 2024]

Ministerio de Hidrocarburos y Energías (2023b). YLB se abre paso en la industria del litio y apunta a captar nuevas inversions. *Press release.* Online: https://www.mhe.gob.bo/2023/12/27/ylb-se-abre-paso-en-la-industria-de-litio-y-apunta-a-captar-nuevas-inversiones/ [accessed Jan 23, 2024]

Ministerio de Hidrocarburos y Energías (2024). YLB lanza segunda convocatoria internacional para nuevos proyectos de recursos evaporíticos en 7 salares de Potosí y Oruro. *Press release.* Online: https://www.mhe.gob.bo/2024/01/26/ylb-lanza-segunda-convocatoria-internacional-para-nuevos-proyectos-de-recursos-evaporiticos-en-7-salares-de-potosi-y-oruro/ [accessed Jan 23, 2024]

Obaya, M. (2021). The evolution of resource nationalism: The case of Bolivian lithium. In: *The Extractive Industries and Society*, 8(3), article 100932. https://doi.org/10.1016/j.exis.2021.100932

Parsons, T. (2013). Indigenous Political Representation in Bolivia. *Dissertation at the University of Missouri.* Online: https://mospace.umsystem.edu/xmlui/handle/10355/37799 [accessed Jan 24, 2024]

Rodriguez Fernandez, G. V. (2020). Neo-extractivism, the Bolivian state, and indigenous peasant women's struggles for water in the Altiplano. In: *Human Geography*, 13(1), 27–39. https://doi.org/10.1177/1942778620910896

Sanchez-Lopez, M. D. (2019), From a White Desert to the Largest World Deposit of Lithium: Symbolic Meanings and Materialities of the Uyuni Salt Flat in Bolivia. In: *Antipode*, 51, 1318-1339. https://doi.org/10.1111/anti.12539

Svampa, M. & Stefanoni, P. (2007). Bolivia: memoria, insurgencia y movimientos sociales. *Collección Resistencias y Alternativas*. Buenos Aires: El Colectivo, Clasco. Online: https://libreria.clacso.org/publicacion.php?p=78&c=6

World Commission on Environment and Development (1987). Brundtland Report. Our Common Future. Online: https://www.are.admin.ch/are/en/home/media/publications/sustainable-development/brundtland-report.html [accessed Jan 25, 2024]

YLB (2017). 26 Empresas Presentaron sus Expresiones de Interés. Online: https://www.ylb.gob.bo/archivos/notas_archivos/04.pdf [accessed Jan 19, 2024]

YLB (2020). Nuevo gerente de empresa estatal de litio Bolivia dice producción metal deberá ser local. Online: https://www.ylb.gob.bo/archivos/notas_archivos/entrevistareuters.pdf [accessed Jan 27, 2024]

YLB (2023a). YLB y Altmin de la India firman convenio de desarrollo de materiales activos para baterías de litio. Online: https://www.ylb.gob.bo/resources/img/10112023.pdf_[accessed Jan 19, 2024]

YLB (2023b). India expresa interés en producer cátodos de baterías con litio boliviano. Online: https://www.ylb.gob.bo/resources/img/26102023.pdf [accessed Jan 19, 2024]

YLB (2024). YLB pone a prueba tecnología de extracción directa de litio ofertada por empresas de Rusia y China. Online: https://www.ylb.gob.bo/resources/img/21012024.pdf_[accessed Jan 19, 2024]

YOUR KNOWLEDGE HAS VALUE

- We will publish your bachelor's and
  master's thesis, essays and papers

- Your own eBook and book -
  sold worldwide in all relevant shops

- Earn money with each sale

Upload your text at www.GRIN.com
and publish for free